Tamika K. Fordham

Tamika K. Fordham

ISBN-13:
978-1981314287

ISBN-10:
1981314288

DEDICATION

This book is dedicated to my past and present students. Always remember what inspire you to learn and to go above and beyond what is required. Keep your imagination open, be creative, solve problems, engage in the process, and always reflect.. The future is in your hands and with my help I know you are college ready.

ACKNOWLEDGMENTS

I want to express my gratitude to my family, especially my mother (Eartha Carson), sister (Michelle Thompson) and husband (Kendrick Fordham), my brother (Marion Kelley), and my mother in law (Shirley Fordham),Dorchester County School District Four and Calhoun County Public Schools for providing me with the tools needed to be successful.

As the sky darkens, the moon shines brighter and brighter in the night sky. The Moon reflects light from the Sun and just like Earth, half of the Moon is always lit by the Sun. Because of the positions of the Sun, the Moon appears to change shape. The amount of reflected light from the Moon that is seen from Earth determines its phase. The craters found on the moon's surface are caused by asteroids and comets when they crash onto the surface of the moon.

Essential Question:
Analyze and interpret data from observations to describe patterns in the location, movement, and appearance of the Moon.

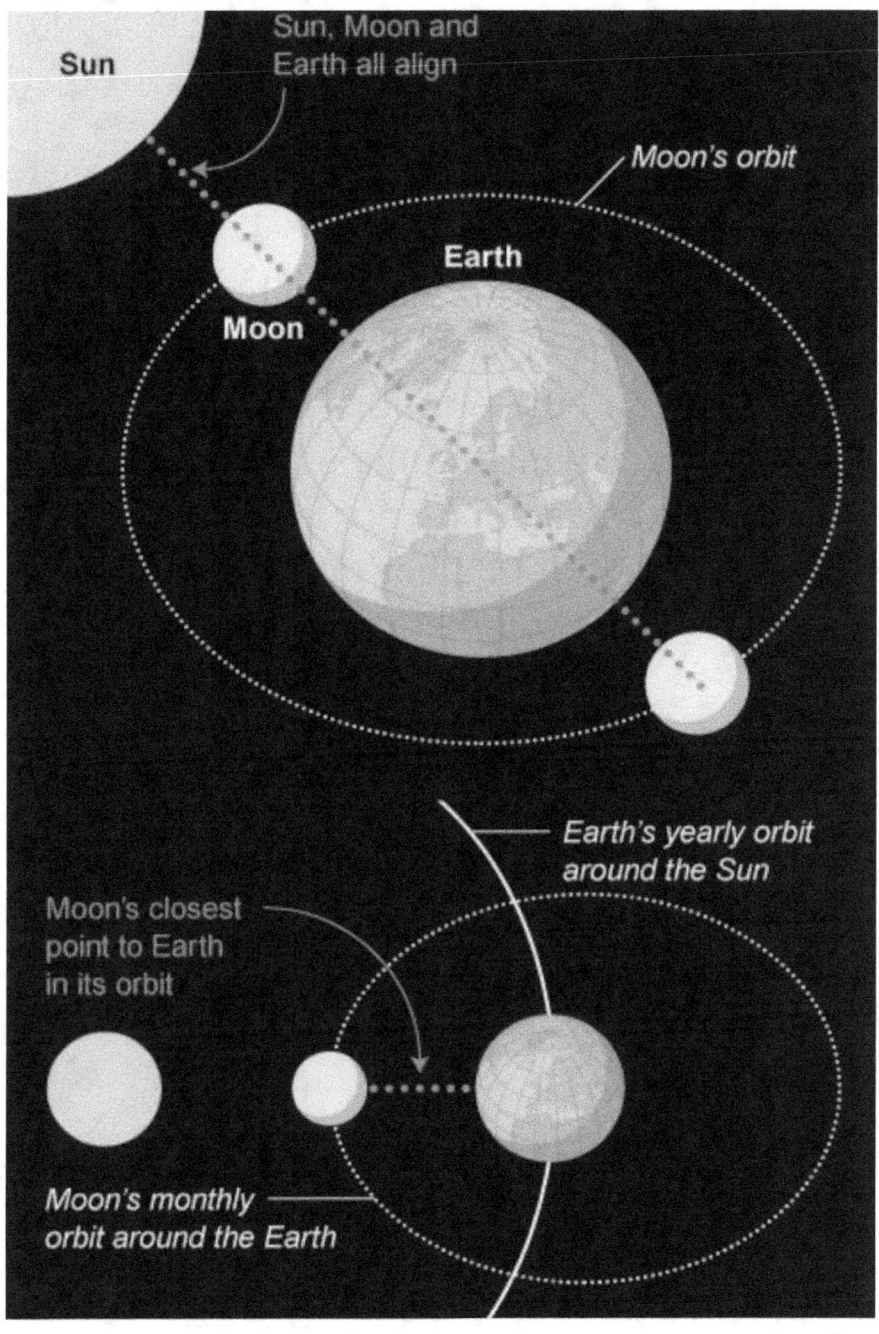

We see different amounts of light being reflected on the moon. How much light we see depends on the positions of Earth, moon and sun. The changing shapes of the Moon are called *phases.* There are four main phases of the moon. They are; New Moon, Quarter Moon, Full Moon, and Crescent Moon.

Essential Questions:

1. Look up at the Moon in the night sky. How do you think it will change over two weeks?

2. Draw a picture and label the phase of the moon in the space below.

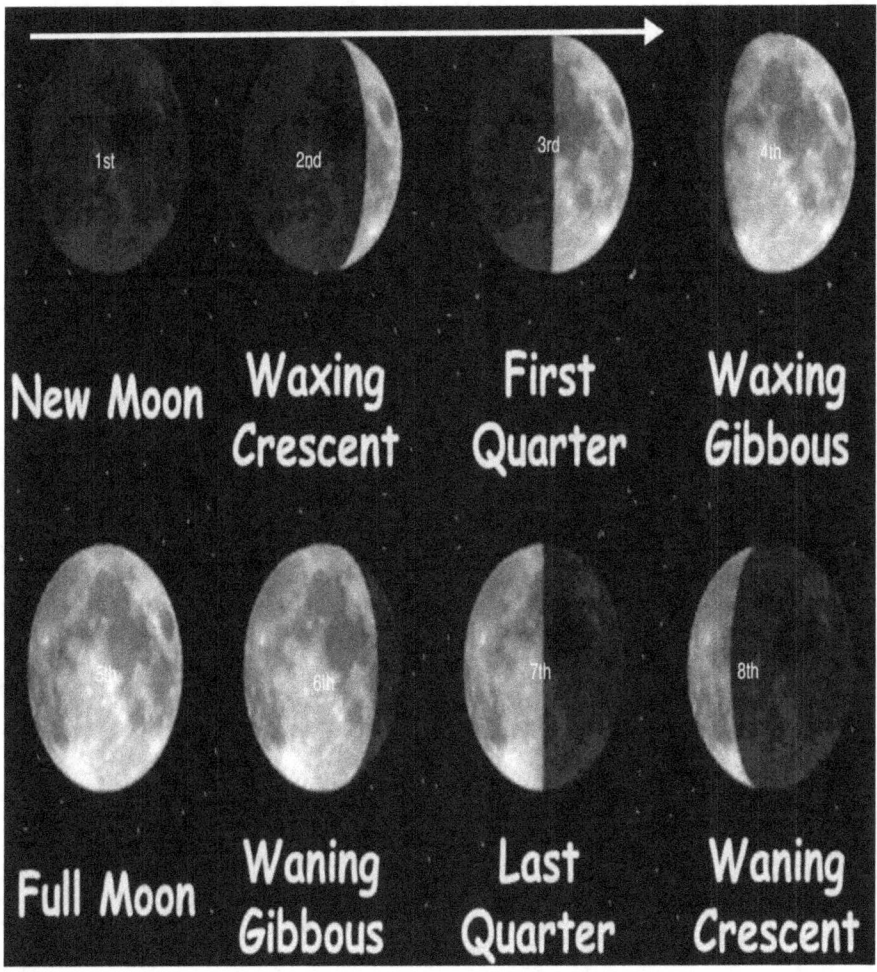

During a **new moon** the entire half/side of the Moon facing Earth is dark. The moon looks dark because no light is being reflected. The moons position is directly between the sun and Earth.

During a *quarter moon* – half of the side of the Moon facing Earth is lit and the other half is dark. The Moon appears as a half circle; there are two quarter-moon phases in the cycle. When the moon is a quarter of its way around Earth , it is waxing or growing on the right. It is in its *first quarter.* It is sometimes called a half moon. When the moon is three quarters of its way around Earth. We see it half lit again. It is waning on the left side and we called it *third quarter* or half moon.

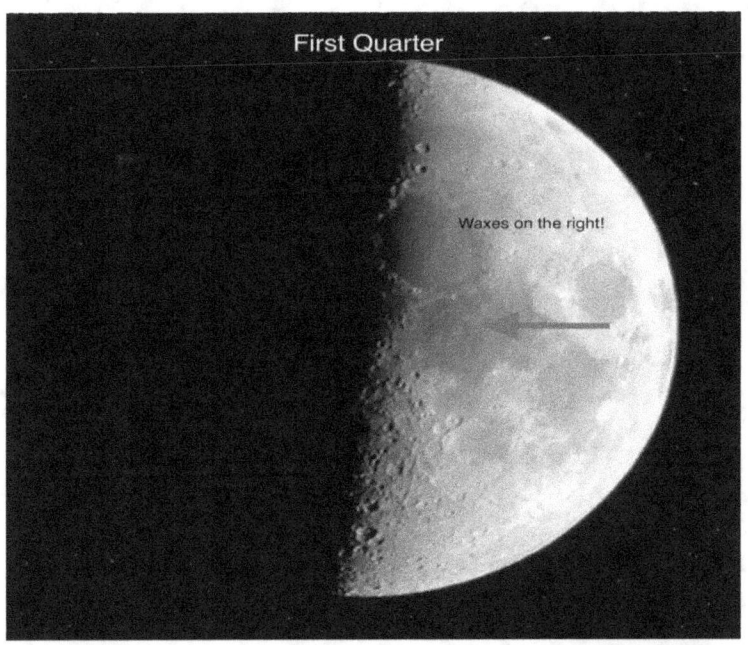

First Quarter

Waxes on the right!

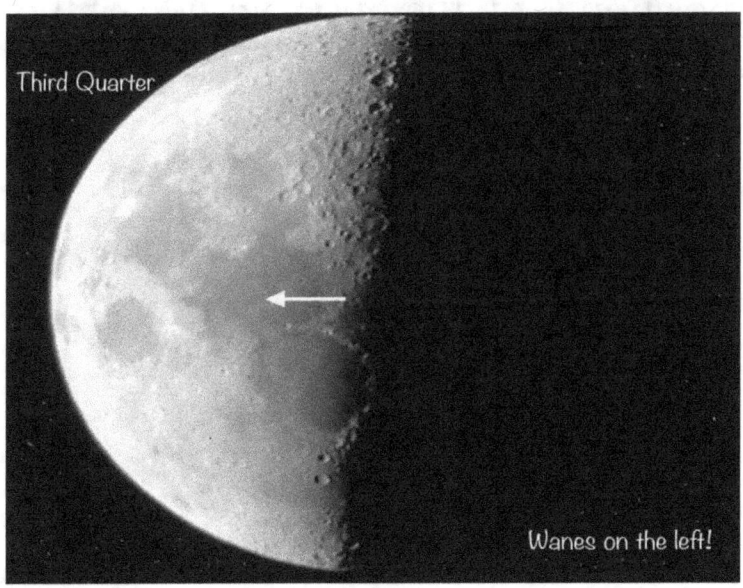

Third Quarter

Wanes on the left!

About two weeks after the new moon will appear a *Full moon* . The entire half/side of the Moon facing Earth is lit and the Moon appears as a full circle. The entire face of the moon shines bright in the night sky. The moons position is on the opposite side of the sun and Earth.

In a few days a *Crescent* shape is lit on the moon. A small section (less than a quarter moon) of the half/side of the Moon facing Earth is lit. When the moon is waxing, the lit surface we see is growing or getting bigger on the right side. When the moon is waning, it is shrinking on the left side. The waning crescent is the last phase of the moon.

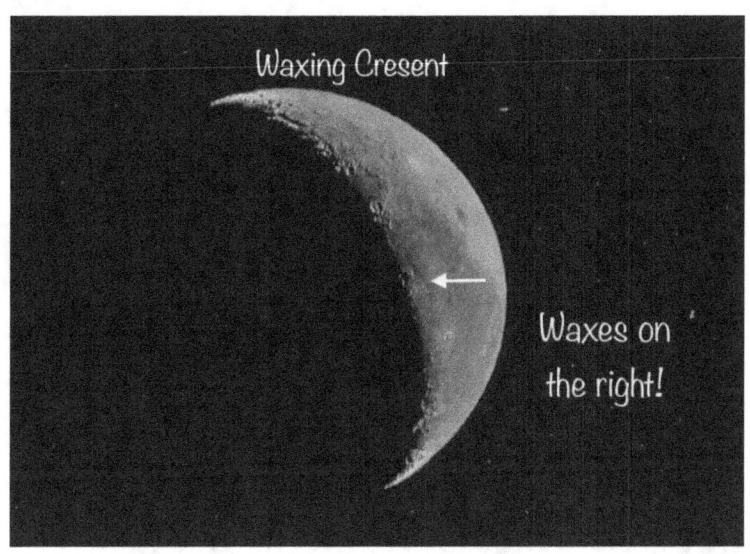

The change in the Moon's phases from new moon to new moon takes about one month, four weeks, or 29½ days. The moon is not a planet, but a satellite of the Earth. The surface area of the moon is 14,658,000 square miles or 9.4 billion acres. Only 59% of the moon's surface is visible from earth. The moon rotates at 10 miles per hour compared to the earth's rotation of 1000 miles per hour. Temperatures on the moon range from 200° F (123° C) during the day to -300° F (-233° C) at night.

Essential Question:

Which planet in our solar system has a surface and temperatures similar to the moon?

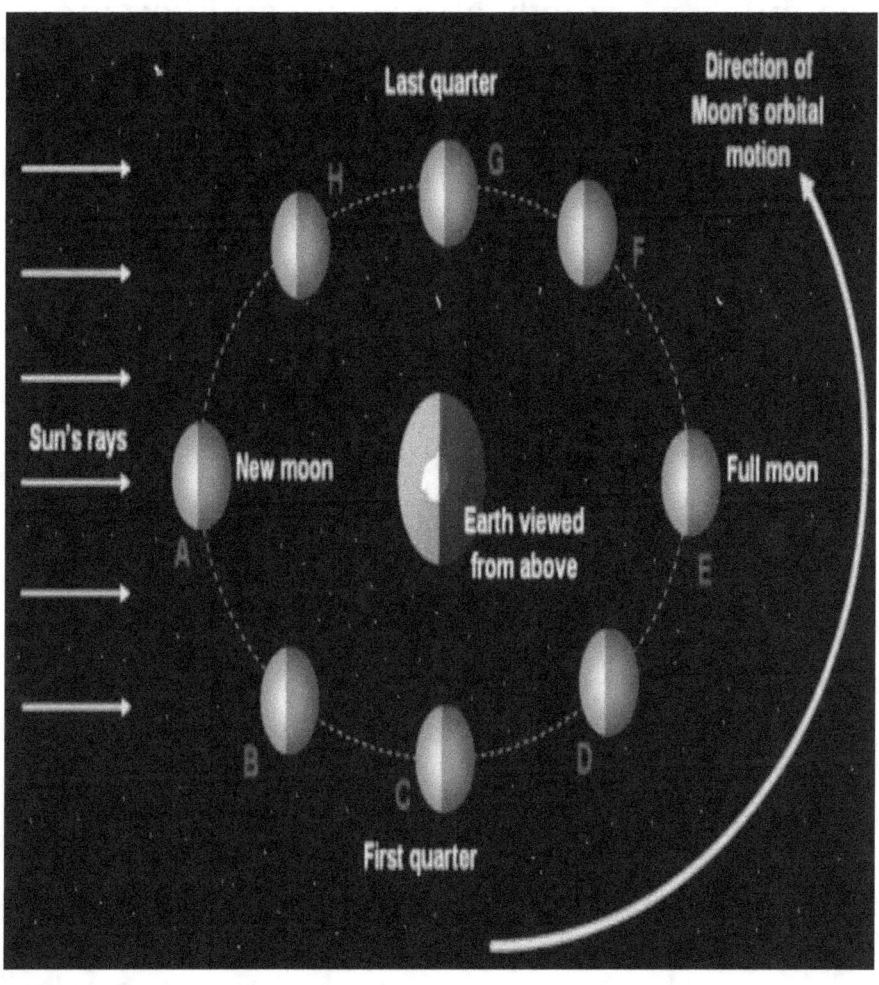

The gravitational relationship between the sun, Earth and moon causes tides. The moon's gravity is 1/6 that of Earth. Its pull on Earth is strong enough to cause tides. When the moon is directly above you and you are along the coast, you will encounter a high tide. Tides are periodic rises and falls of large bodies of water. The gravitational attraction of the moon causes the oceans to bulge out in the direction of the moon. A *Spring tide* is a rare high tide that occurs during the new moon and full moon phase. Neap tide or low tide occur during quarter moons. Another bulge occurs on the opposite side, since the Earth is also being pulled toward the moon (and away from the water on the far side). Earth experiences two tides each day (high and low). Isaac Newton was the first person to explain tides scientifically.

Essential Question:

Who invented the first telescope? What is it used for and explain how it works?

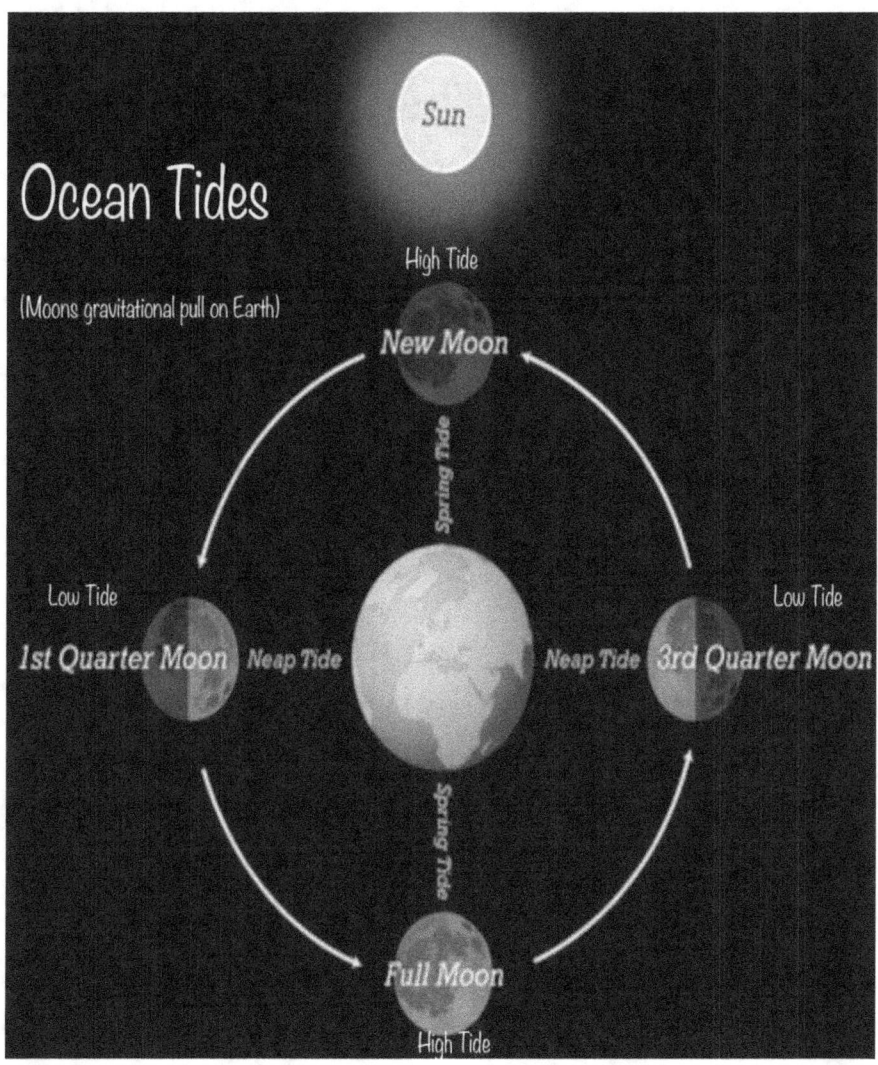

"Lunar" comes from "Luna" the Latin word for moon. A lunar eclipse can last for three hours. *Lunar eclipses* occurs when the Earth is located in between the Sun and the Moon. When the sun, Earth, and moon are exactly lined up, a lunar eclipse when happen. The Earth blocks the sunlight that the moon reflects and cast a shadow on the moon. The moon will have a reddish glow instead of its usual white color.

Solar eclipses occurs when the Moon is between the Sun and Earth. When the moon's orbit or revolution is not tilted, and The moon lines up exactly between the sun and Earth, a solar eclipse happens. The moon shadow will be cast on Earth. This can last up to seven hours.

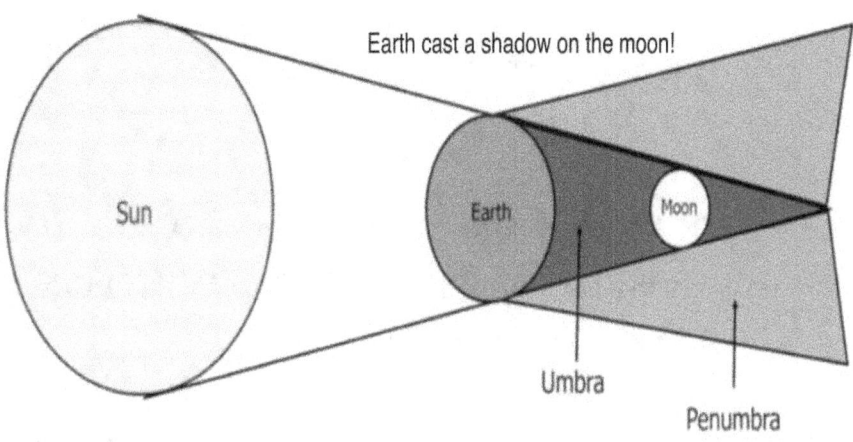

Earth cast a shadow on the moon!

Sun

Earth

Moon

Umbra

Penumbra

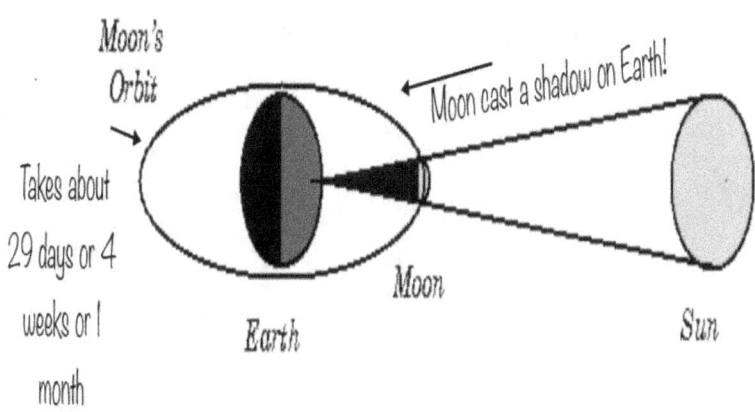

Moon's Orbit

Takes about 29 days or 4 weeks or 1 month

Moon cast a shadow on Earth!

Earth

Moon

Sun

MOON FUN FACTS!

1. THE MOON DOES NOT HAVE AN ATMOSPHERE; THERE IS NO WIND ON THE MOON.

2. NEIL ARMSTRONG BECAME THE FIRST MAN TO WALK ON THE MOON.

3. APOLLO 11 WAS THE AMERICAN SPACE MISSION TO FIRST REACH AND LAND ON THE MOON.

Teach Me the Faces of the Moon

ABOUT THE AUTHOR

My name is Tamika K. Fordham, wife of Kendrick Fordham and mother of Kennedy Fordham. I was selected the 2017 NAACP Teacher of the Year, 2016-2017 District Teacher of the Year for Calhoun County Public Schools and as 2017 STAR Teacher.

I graduated with honors from South Carolina State University and earned my Masters of Arts in Teaching in Early Childhood. I also graduated with honors from Liberty University and received my Educational Specialist Degree in Leadership. I served in the United States Army Reserve throughout my college years and most of my teaching career. I became a highly qualified early childhood generalist teacher when I earned my National Board Certification. I currently teach science and have taught all subjects!

I give all of my praise to my God!

www.ingramcontent.com/pod-product-compliance
Lightning Source LLC
Chambersburg PA
CBHW030041230526
45472CB00002B/627